CONTENTS

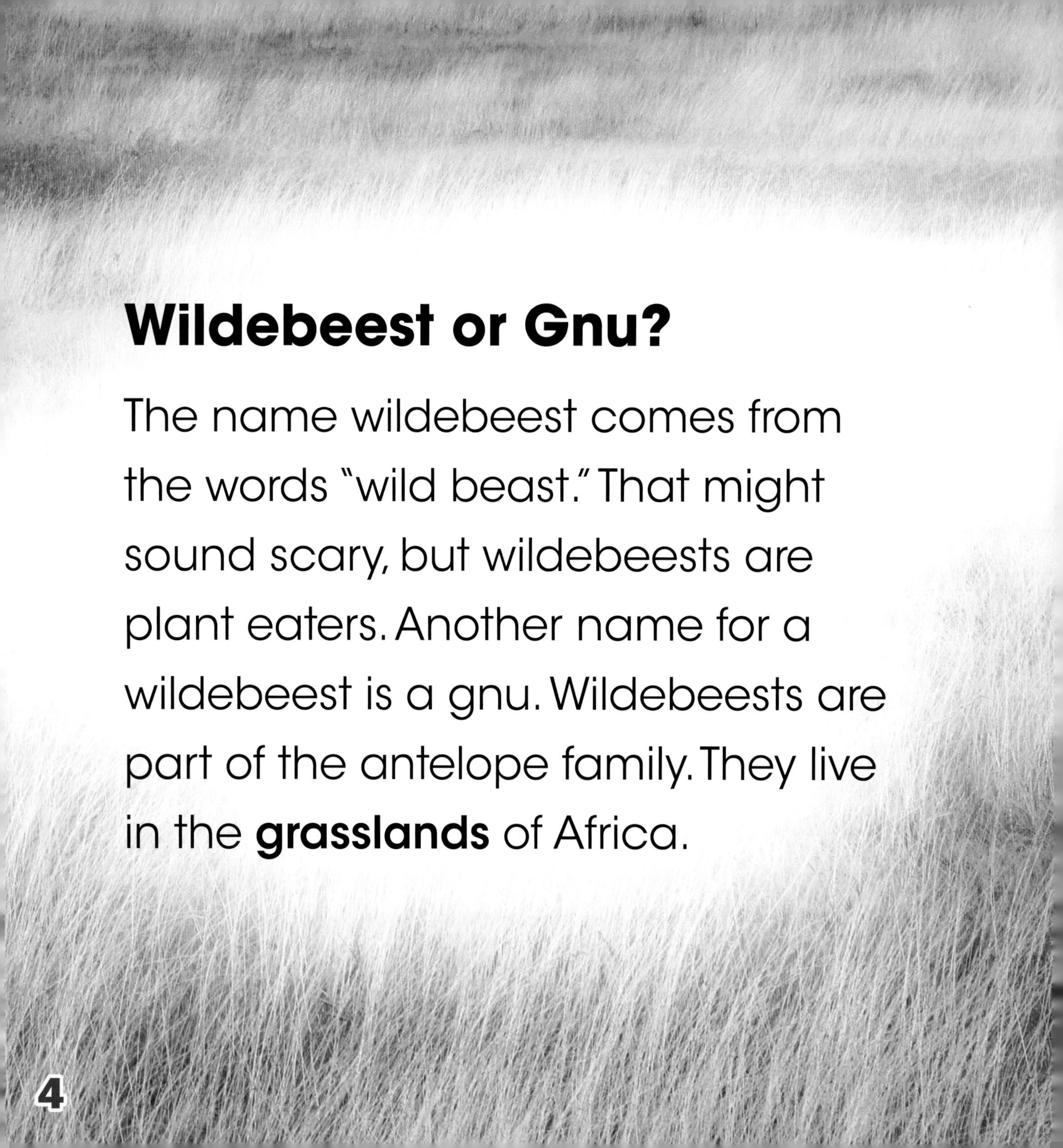

Wildebeest or Gnu?

The name wildebeest comes from the words "wild beast." That might sound scary, but wildebeests are plant eaters. Another name for a wildebeest is a gnu. Wildebeests are part of the antelope family. They live in the **grasslands** of Africa.

What Do They Look Like?

Wildebeests are large animals with big heads, thick necks, and thin legs. The short hair on their body may be dark gray or brown in color. They also have longer, darker manes, pointed beards, and tails. Both **males** and **females** have horns.

Blue, Black, and White Bearded

There are different kinds of wildebeests. The blue wildebeest is the largest kind. It weighs up to 600 pounds (272 kg). Western white-bearded wildebeests and black wildebeests are much smaller and shorter. Black wildebeests weigh up to 323 pounds (147 kg).

western white-bearded wildebeest

black wildebeest

blue wildebeest

Herd Life

Wildebeests live in groups called herds. Females usually stay with the same herd for life. Male wildebeests are called bulls. They leave the herd when they're about a year old. Bulls may fight with each other over **territory** and **mates**.

There may be as many as 1,000 wildebeests in a herd. Living in a herd helps wildebeests stay safe. Top predators such as hyenas, lions, and cheetahs hunt wildebeests. When female wildebeests give birth, they do so in the middle of the herd.

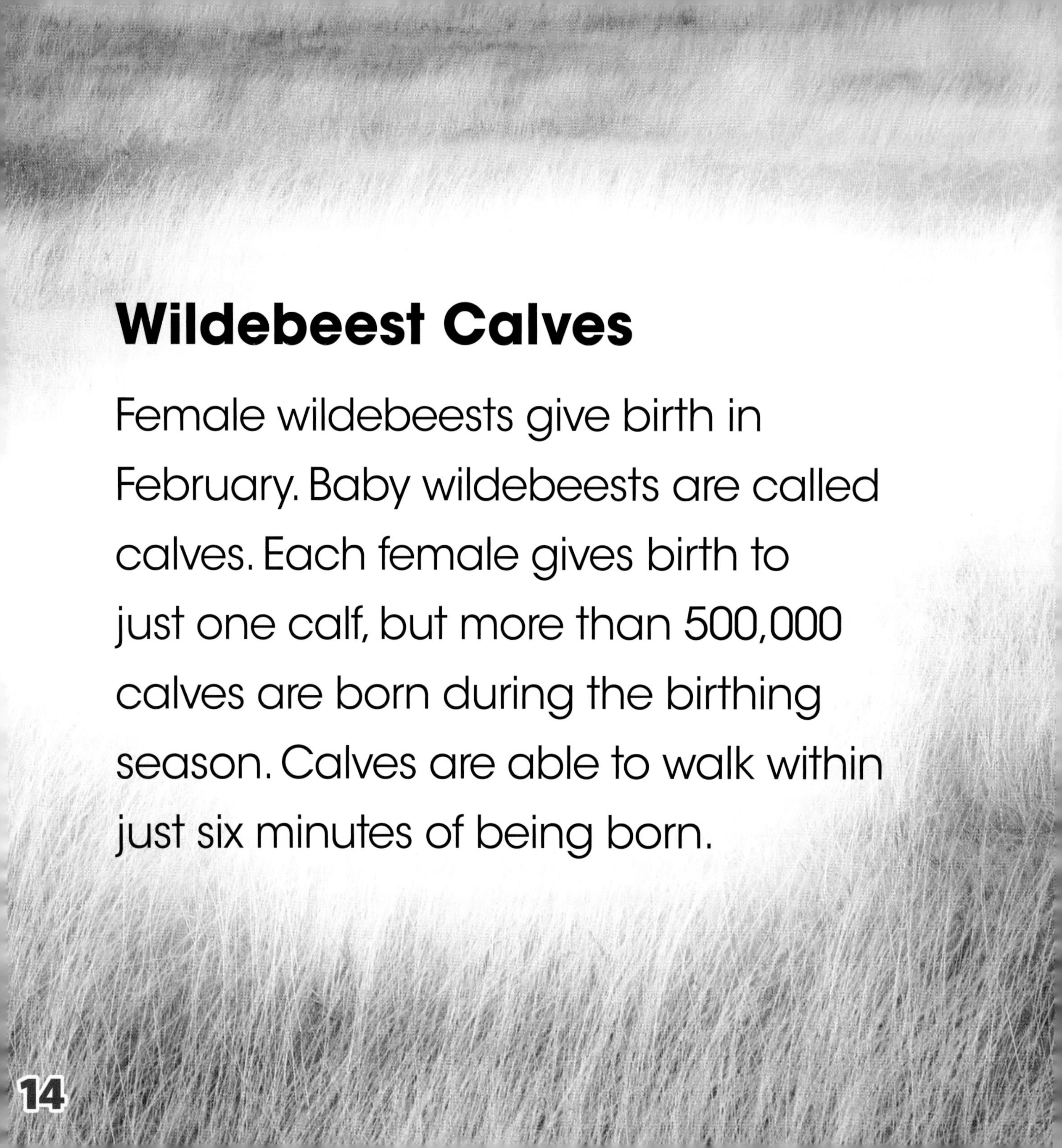

Wildebeest Calves

Female wildebeests give birth in February. Baby wildebeests are called calves. Each female gives birth to just one calf, but more than 500,000 calves are born during the birthing season. Calves are able to walk within just six minutes of being born.

Hungry Beasts

Wildebeests spend most of their time looking for food and water. They eat all day and night. Wildebeests eat plants and bushes but enjoy grasses most. Their teeth are specially made to rip and chew up grass.

On the Move

Wildebeest herds migrate, or move from one place to another, seasonally. They follow the rain, looking for fresh grass and water. A group of wildebeests from the **Serengeti** migrates in a circle of 500 to 1,000 miles (805 to 1,610 km) each year.

The great wildebeest migration takes place every year, usually during May or June. Nearly two million wildebeests make the journey! It's widely considered one of the wonders of the natural world and many people travel to Africa to watch it.

Fancy Features

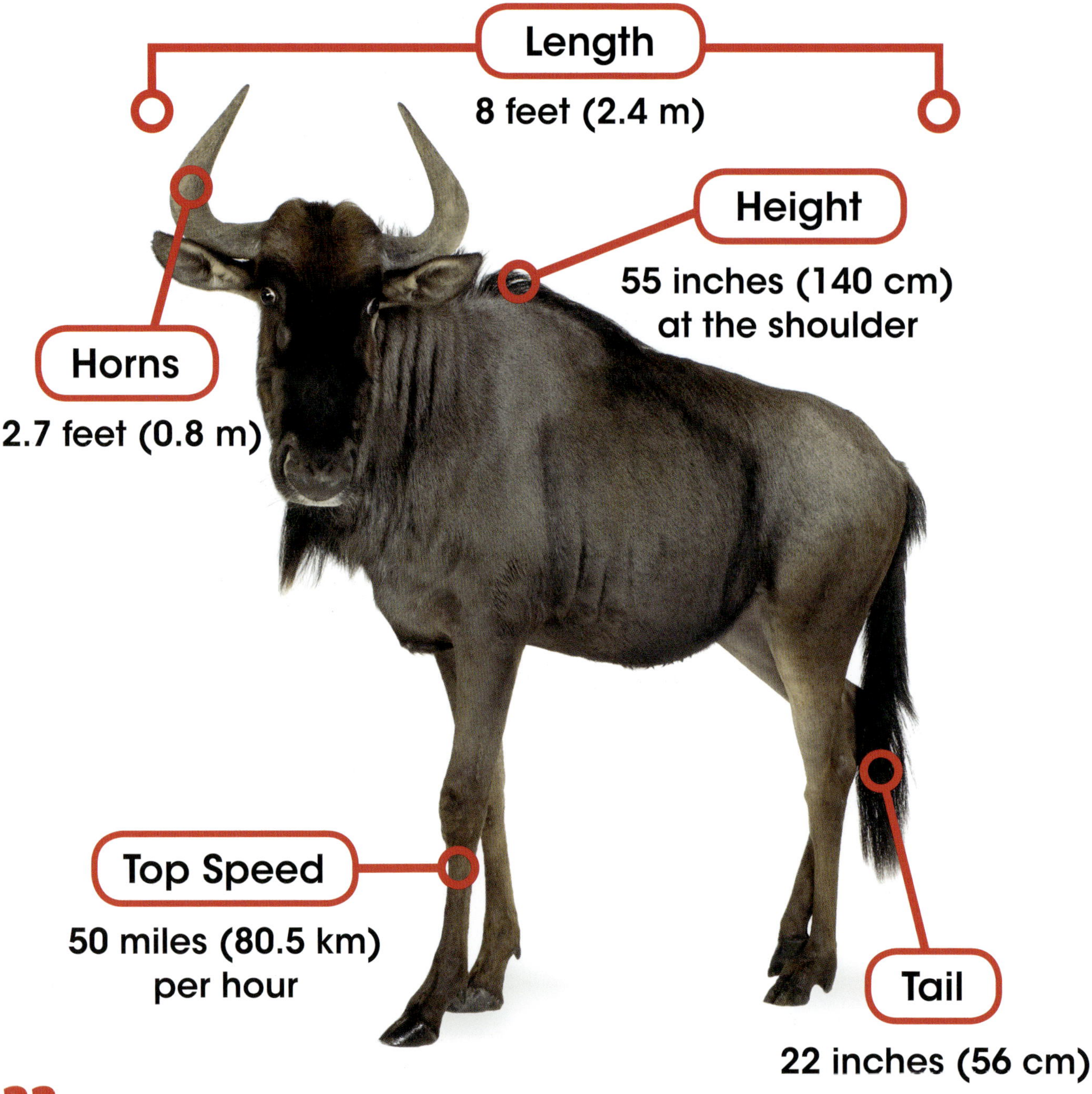

GLOSSARY

females: Girls.

grasslands: Land on which grass is the main plant life.

males: Boys.

mates: One of two animals that come together to produce babies.

Serengeti: A region, or place, in Tanzania and part of Kenya.

territory: Land or space that is protected by an animal for its use.

INDEX

WEBSITES

Due to the changing nature of Internet links, PowerKids Press has developed an online list of websites related to the subject of this book. This site is updated regularly. Please use this link to access the list: www.powerkidslinks.com/aotg/wildebeest

ANIMALS OF THE GRASSLANDS

WANDERING WILDEBEESTS

THERESA EMMINIZER

PowerKiDS press.
New York

Published in 2021 by The Rosen Publishing Group, Inc.
29 East 21st Street, New York, NY 10010

First Edition

Portions of this work were originally authored by Mary Molly Shea and published as *Wildebeests*. All new material this edition authored by Theresa Emminizer.

Editor: Theresa Emminizer
Book Design: Michael Flynn

Photo Credits: Cover, p. 1 Jez Bennett/Shutterstock.com; p. 5 nwdph/Shutterstock.com; p. 7 paranyu pithayarungsarit/Moment/Getty Images; p. 9 (Western white-bearded wildebeest) Lou Coetzer/Image Source/Getty Images; p. 9 (blue wildebeest) Simon Eeman/Shutterstock.com; p. 9 (black wildebeest) Roy Coetzee/Shutterstock.com; p. 11 EcoPrint/Shutterstock.com; p. 13 javarman/Shutterstock.com; p. 15 Marcelle Robbins/Shutterstock.com; p. 17 Marc Guitard/Moment/Getty Images; p. 19 Gudkov Andrey/Shutterstock.com; p. 21 Darrell Gulin/The Image Bank/Getty Images; p. 22 Eric Isselee/Shutterstock.com.

Cataloging-in-Publication Data

Names: Emminizer, Theresa.
Title: Wandering wildebeests / Theresa Emminizer.
Description: New York : PowerKids Press, 2021. | Series: Animals of the grasslands | Includes glossary and index.
Identifiers: ISBN 9781725317147 (pbk.) | ISBN 9781725317161 (library bound) | ISBN 9781725317154 (6 pack)
Subjects: LCSH: Gnus--Juvenile literature.
Classification: LCC QL737.U53 E44 2021 | DDC 599.64'59--dc23

Manufactured in the United States of America

CPSIA Compliance Information: Batch #CSPK20. For Further Information contact Rosen Publishing, New York, New York at 1-800-237-9932.